Computer Mathematics

Sets, Numbers and Flowcharts

Introduction to Logic

Computer Number Bases

Boolean Algebra and Switching Circuits

ALSO BY WILLIAM R. PARKS

Paperback editions

Beginning Algebra

Introduction to Statistics and Probability

Computer Mathematics

Sets, Numbers and Flowcharts

Introduction to Logic

Computer Number Bases

Boolean Algebra and Switching Circuits

William R. Parks, B.S., M.S., Ed.M.

Hershey
Books

ISBN: 978-0-88493-065-5

CONTENTS

SETS AND NUMBERS
SETS AND FLOWCHARTS

These two sections cover the fundamental concepts of sets, numbers and flowcharts used in data processing and computer applications.

The material is presented in a simple and easy to understand format. We utilize a modified form of programmed instruction (PI) in all of the presentations for various topics in this book. Small bits of information are presented in each frame before advancing to the next frame. Exercises are listed after several frames, followed by an answer key.

We also list example problems followed by the solution which is given right after the problem.

The immediate answer feedback approach to problem solving is important when learning new concepts. The exercises serve the same purpose. Answers are given right after each set of exercises.

Sets and numbers go hand and hand with learning basic concepts of flowcharting The mathematics of set theory is used to explain the basic principles of flowcharting for data processing applications.

A flowchart is a diagram that shows various types of boxes connected with lines and arrows which show the steps being taken and decisions being made within a computer or data processing application. Flowcharts are prepared before programs are written. Flowcharts may also be used in business reports to explain work-flows.

SETS AND NUMBERS

1. A set is a well-defined collection of things conceived as a whole.

well-defined

EXAMPLES:
(a) the set of tools in a tool chest
(b) the set of playing cards
(c) the set of even numbers 2, 4, 6, and 8
(d) the set of customer names in a company file

- - - - - - - - - - - - -

2. The things in a set are called either members or elements of the set.

EXAMPLES:
(a) 3 is an element of the set of counting numbers.
(b) The set of vowels contains five elements: a, e, i, o, u.
(c) A record of one customer may be an element of the set of accounts receivable.
(d) The month of October is a member of the set of months having exactly 31 days.

- - - - - - - - - - - - -

3. It is customary to name a set by a capital letter, and there are two ways to define a set: by a word description written between braces { }, and by listing the elements between braces { }.

EXAMPLE:
D = { all decimal digits }
= { 0, 1, 2, 3, 4, 5, 6, 7, 8, 9 }

- - - - - - - - - - - - -

4. It is customary to use small letters to designate elements in a set. If a small letter is used with braces and a vertical bar "|", this is called set-builder notation. The vertical bar stands for the verbal expression "such that."

EXAMPLE:
The symbolic statement $A = \{n \mid n \text{ is a counting number less than } 10\}$ is read, "A is the set of all elements n such that n is a counting number less than 10." This same set can be defined by a listing or by a word description. $A = \{1, 2, 3, 4, 5, 6, 7, 8, 9\}$
$= \{\text{counting numbers less than } 10\}$

- - - - - - - - - - - -

5. The Greek letter "∈" (epsilon) stands for the phrase, "is an element of," or the phrase, "is a member of."
If $B = \{1, 3, 5, 7, 9\}$, then the statement "five is an element of the set B" is abbreviated in set notation by "$5 \in B$."
The statement "6 is not an element of set B" is written in set notation as "$6 \notin B$."

EXAMPLES:
(a) If $V = \{\text{vowels of the alphabet}\}$, then $i \in V$ and $j \notin V$.
(b) If $E = \{\text{even counting numbers}\}$, then $6 \in E$ and $7 \notin E$.

- - - - - - - - - - - -

Note: Set theory notation was invented for the purpose of discussing collections of similar things. It is also a very useful methodology for defining business and mathematical operations. Georg Cantor (1845 - 1918) a German Mathematician invented set theory. Today it is used in mathematics, business, data processing and computer science.

6. The universal set is a set of elements under discussion at a particular point in time.

EXAMPLE:
Let the universal set $U = \{\text{counting numbers}\}$
$= \{1, 2, 3, 4, \ldots\}$

If you were asked to find the solution to each of the following equations within the given universal set, then your solutions would be stated as follows:

(a) $6 \div 2 = n$. Here the solution is $n = 3$.
(b) $6 \div 3 = n$. Here the solution is $n = 2$.
(c) $6 \div 4 = n$. Here the solution does not exist. The apparent solution $n = \frac{3}{2}$ is disqualified since the universal set does not contain the element $\frac{3}{2}$. Therefore, there is no solution in EXAMPLE (c).

- - - - - - - - - - - -

Note: A universal set is the total collection of all objects under discussion in a particular business or mathematical application. Objects in a universal set are considered members of the universal set. Subsets of the universal set can be made up of objects from the universal set. We will take up subsets in the next section.

7. EXERCISE 1

Answer all questions before checking the answer key in Frame 8.

1. A well-defined collection of things conceived as a whole is called a ________________________.

2. ____________ letters are used to name sets and ________ letters are used to designate elements in a set.

3. The set of all pretty girls is well defined. True or False?

   ____

4. Define the set of counting numbers less than seven by listing all elements. C = ________________________________.

5. Define the set in Question 4 by a word description.
   C = ____.

6. Define the set in Question 4 using set-builder notation.
   C = ____.

7. Define the set of all counting numbers in set-builder notation.
   N = ____.

8. Given $E = \{2, 4, 6, 8, 10, \ldots\}$ (the three dots mean the sequence is infinite), determine which statements are true and which are false:
   (a) $6 \in E$
   (b) $25 \notin E$
   (c) $30 \notin E$
   (d) $51 \in E$

9. If the universal set of elements under discussion is the set of counting numbers, then what is the solution to each equation:
   (a) $9 \div 3 = n$ $\quad n =$ ____________
   (b) $10 \div 3 = n$ $\quad n =$ ____________
   (c) $5 \div 8 = n$ $\quad n =$ ____________
   (d) $30 \div 6 = n$ $\quad n =$ ____________

8. ANSWERS FOR EXERCISE 1

1. set

2. Capital, small

3. False, since there is no objective criterion by which one could judge prettiness.

4. $\{1, 2, 3, 4, 5, 6\}$

5. {counting numbers less than 7}

6. {n | n is a counting number less than 7}

7. {n | n is a counting number}

8. (a) true (b) true (c) false (d) false

9. (a) $n = 3$ (b) no solution (c) no solution (d) $n = 5$

9. A set that is contained in another set is called a <u>subset</u>. The symbol "⊆" stands for the phrase, "is a subset of."

EXAMPLES:

(a) If M = { a, b, c, d, e } and N = { a, c, e } then N ⊆ M.

N⊆M

(b) If A = { 1, 2, 3, 4, 5 } and B = { 1, 3, 5 } then B ⊆ A.

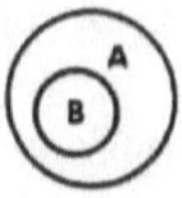

B⊆A

NOTE: We can also say that A is a subset of itself or "A ⊆ A."

10. The <u>empty set</u> is the set that has no elements. The symbol for the empty set is either the Greek letter "∅" (pronounced "phi") or braces { } with no elements contained in them.

EXAMPLES:

(a) The set of women who were presidents of the United States is the empty set.
(b) The set of odd numbers between 3 and 5 is the empty set.
(c) The empty set is a subset of every set.
(d) The set of counting numbers between 2 and 3 is the empty set.

Note: Another name for "empty set" is "null set." The number of elements in a null set is zero. The size of a set is called its cardinality and is found by counting the elements in the set.

11. In the study of set theory it is convenient to distinguish between two kinds of sets: finite and infinite sets. In Frame 5, V, the set of vowels of the alphabet, is an example of a finite set. In the same Frame, E, the set of even counting numbers, is an example of an infinite set.

EXAMPLES:

(a) An infinite set is a set that contains an inexhaustible number of elements.
$C = \{n | n \text{ is a counting number}\}$
$= \{1, 2, 3, 4, 5, 6, 7, \ldots\}$

(b) A finite set is a set whose total number of elements can be expressed by a whole number. $D = \{1, 5, 9, 13\}$. This set is finite. It contains exactly four elements: 1, 5, 9, and 13.

- - - - - - - - - - - - -

12. Two sets A and B are said to be equal if A and B have the same elements. If two sets do not have the same elements, then they are unequal.

EXAMPLES:

(a) If $A = \{2, 4, 6\}$ and $B = \{4, 2, 6\}$, then $A = B$.
(b) If $C = \{1, 2, 3\}$ and $D = \{2, 3, 4\}$, then $C \neq D$.

- - - - - - - - - - - - -

Note: Set theory is a branch of mathematics which can be used to describe most concepts in mathematics. Although sets can be applied to collections of many types of things, set theory is mainly applied to mathematics and computer applications which we shall study later in the text.

13. Two sets A and B are said to be equivalent if A and B have the same number of elements. It must be possible to match elements or make a one-to-one correspondence between the elements of each set without having an extra element left over in either set. The symbol for equivalence is "$\leftrightarrow$."

EXAMPLES:

(a) If $A = \{a, b, c, d\}$ and $B = \{1, 2, 3, 4\}$, then $A \leftrightarrow B$. The symbolic statement "$A \leftrightarrow B$" is read, "set A is equivalent to set B." The elements of each set can be placed in a one-to-one correspondence.

$$\begin{array}{rcccccl} A = \{ & a, & b, & c, & d & \} \\ & \updownarrow & \updownarrow & \updownarrow & \updownarrow & \\ B = \{ & 1, & 2, & 3, & 4 & \} \end{array}$$

(b) If $E = \{2, 4, 6, 8, \ldots\}$ and $O = \{1, 3, 5, 7, \ldots\}$, then $E \leftrightarrow O$. In this case the two sets E and O are infinite and equivalent because their elements can be placed in a one-to-one correspondence.

$$\begin{array}{rccccccccccl} E = \{ & 2, & 4, & 6, & 8, & 10, & 12, & 14, & \ldots & 100, & \ldots & \} \\ & \updownarrow & \updownarrow & \updownarrow & \updownarrow & \updownarrow & \updownarrow & \updownarrow & & \updownarrow & & \\ O = \{ & 1, & 3, & 5, & 7, & 9, & 11, & 13, & \ldots & 99, & \ldots & \} \end{array}$$

- - - - - - - - - - - -

Note: The idea of sets is useful in many areas of communication not just with numbers in mathematics or in data processing. We can make lists of favorite songs. In business, the idea of sets is useful in categorizing products or items needed in the manufacture of certain goods. Sets are useful when deciding what belongs in a specific category or to identify a set of specific data under consideration.

14. EXERCISE 2

Determine which statements are truw and which are false, given the following four sets: A, B, C and D. Answer all questions before checking the answer key in Frame 15.

$A = \{1, 2, 5, 7\}$
$B = \{5, 2, 7, 1\}$
$C = \{5, 10, 15, 20\}$
$D = \{5, 10, 15\}$

1. $10 \in C$
2. $2 \in D$
3. $6 \notin A$
4. $A \leftrightarrow C$
5. $A = B$
6. $A \in B$
7. $D \subseteq C$
8. $B \neq C$
9. $A \leftrightarrow B$
10. $\emptyset \subseteq A$
11. $B \subseteq B$
12. $A \leftrightarrow D$

15. ANSWERS FOR EXERCISE 2

1. true
2. false
3. true
4. true
5. true
6. false
7. true
8. true
9. true
10. true
11. true
12. false

16. If we represent a finite universal set by a rectangle and a smaller subset of the universal set by a circle, then the region outside the circle is called the complement of A and it is written as "$\bar{A}$."

EXAMPLE:
Let the universal set U = { 1, 2, 3, 4, 5, 6, 7, 8, 9 }.
If A = { 2, 4, 6, 8 }, then $\bar{A}$ = { 1, 3, 5, 7, 9 }.

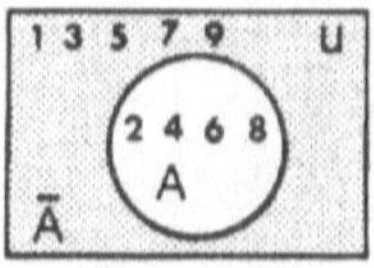

This is called a Venn diagram.

- - - - - - - - - - - - -

17. Two operations that are performed on sets are union and intersection. The outcome of each operation is the formation of a new set.

EXAMPLES:
Let the universal set U = { 1, 2, 3, 4, 5, 6, 7, 8, 9 }.
Let set S = { 1, 2, 3, 4 }
and set R = { 3, 4, 5, 6, 7 }.

(a) The intersection of two sets is the set of all elements common to both sets. The intersection of sets S and R is symbolized by "S ∩ R" (read "S intersection R").
S ∩ R = { 1, 2, 3, 4 } ∩ { 3, 4, 5, 6, 7 } = { 3, 4 }.

Intersection

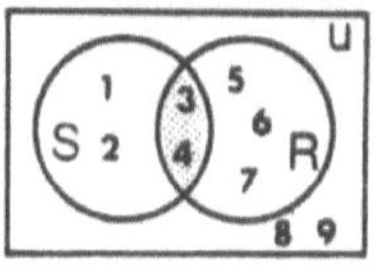

S ∩ R = { 3, 4 }

(Frame 17, contd.)

(b) The <u>union</u> of two sets is a new set which contains all of the elements found in at least one of the two given sets. The union of sets S and R is symbolized by "$S \cup R$" (read "S union R"). $S \cup R = \{1, 2, 3, 4\} \cup \{3, 4, 5, 6, 7\} = \{1, 2, 3, 4, 5, 6, 7\}$.

<u>Union</u>

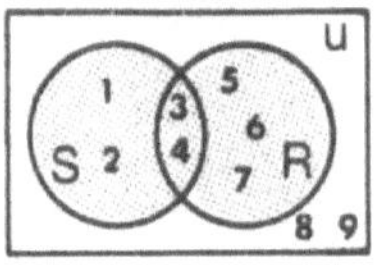

$$S \cup R = \{1, 2, 3, 4, 5, 6, 7\}$$

- - - - - - - - - - - -

Note: In arithmetic, we operate on two numbers when we add them together. This is called a binary operation in arithmetic involving two numbers. We can apply the same concept in set theory by having binary operations involving sets including union, intersection and other types of set operations.

18. EXERCISE 3

Answer all questions before checking the answer key in Frame 19.

Let the universal set $U = \{1, 2, 3, 4, 5, 6, 7, 8, 9\}$.
Answer the following questions:

1. If $A = \{1, 5, 6, 7, 9\}$, then what is $\bar{A}$ (i.e. find the complement of set A by listing its elements). ________

2. Use a Venn diagram to illustrate the relationship between the sets A and $\bar{A}$ within the universal set of Problem 1.

3. If A and $\bar{A}$ are the same as in Problem 1, and $B = \{2, 3, 4, 5, 6, 7\}$, then find each new set:

   (a) $A \cup B =$ ____________ (c) $A \cap \bar{A} =$ ____________

   (b) $A \cap B =$ ____________ (d) $A \cup \bar{A} =$ ____________

- - - - - - - - - - - - -

19. ANSWERS FOR EXERCISE 3

1. $\bar{A} = \{2, 3, 4, 8\}$

2. 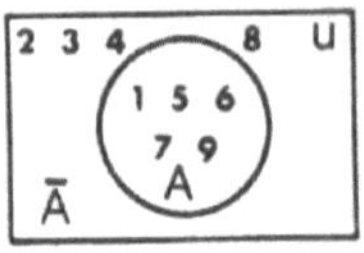

3. (a) $A \cup B = \{1, 2, 3, 4, 5, 6, 7, 9\}$
   (b) $A \cap B = \{5, 6, 7\}$
   (c) $A \cap \bar{A} = \emptyset$ (or empty set)
   (d) $A \cup \bar{A} = U$ (or universal set)

- - - - - - - - - - - -

20. In mathematics there are sets of numbers of particular importance. Given below is a list of important sets. Study these examples.

N = { counting (or natural) numbers } = { 1, 2, 3, 4, 5, 6, . . . }.

W = { whole numbers } = { 0, 1, 2, 3, 4, 5, 6, . . . }.

I = { integers } = { . . . , -3, -2, -1, 0, 1, 2, 3, . . . }.

Q = { rational numbers } = { any number that can be expressed as a ratio of two integers } e.g. $\frac{1}{2}$, $-5\frac{1}{4}$, $\frac{10}{2}$, 7.4, etc.

P = { irrational numbers } = { any number which cannot be expressed as a ratio of two integers } e.g. $\sqrt{2}$, $\sqrt{3}$, $\sqrt{5}$, $\sqrt{10}$, and others whose roots are not perfect.

R = { real numbers } = <u>all the above sets combined into one</u>.

- - - - - - - - - - - - -

21. The following statements are implied by the sets of numbers which are defined in Frame 20.

(a) The set of counting (or natural)numbers is contained in the set of whole numbers and this is contained in the set of integers.

$$N \subseteq W \subseteq I$$

(b) The integers are contained in the set of rational numbers.

$$I \subseteq Q$$

(c) The set of irrational numbers has no elements in common with the set of rational numbers.

$$Q \cap P = \emptyset$$

(d) The collection of all rational numbers and irrational numbers make up the set of real numbers.

$$Q \cup P = R$$

- - - - - - - - - - - - -

22. EXERCISE 4

Answer all questions before checking the answer key in Frame 23.

1. List the elements of the set of integers greater than -1 and less than 5. ______________________

2. Given U = { natural numbers } where U is the universal set. Make a list of elements for each set:
   (a) the set of numbers between 50 and 52 __________
   (b) the set of numbers between 35 and 36 __________
   (c) the set of numbers less than 3 __________

3. Use set-builder notation to designate the following sets:
   (a) the set of natural numbers less than 16 __________
   ____
   (b) the set of integers greater than 10 __________
   ____
   (c) the set of real numbers greater than 3 __________
   ____

4. The set of rational numbers is given in set-builder notation. Write a description of how it is read.

   $Q = \{\frac{a}{b} \mid a \in I,\ b \in I \text{ and } b \neq 0\}$ Note: I = { integers }: see Frame 20. ______________________
   ____

5. Name the set(s) that the following groups of numbers belong to. Use the set names found in Frame 20.
   (a) 1, 2, 5, 7 ______________________
   (b) -1, -5, 6, 12 ______________________
   (c) -2, -5, -8, -25 ______________________
   (d) $\sqrt{5}$, $\sqrt{6}$, $\sqrt{7}$, $\sqrt{20}$ ______________________
   (e) $\sqrt{4}$, $\sqrt{9}$, $\sqrt{16}$, $\sqrt{25}$ ______________________

- - - - - - - - - - - -

23. ANSWERS FOR EXERCISE 4

1. $\{0, 1, 2, 3, 4\}$

2. (a) $\{51\}$
   (b) $\emptyset = \{\ \}$ = the empty set
   (c) $\{1, 2\}$

3. (a) $\{n | n \in N$ and n is less than $16\}$ or
   $\{n | n \in N \text{ and } n < 16\}$
   (b) $\{x | x \in I$ and x is greater than $10\}$ or
   $\{x | x \in I \text{ and } x > 10\}$
   (c) $\{x | x \in R$ and x is greater than $3\}$ or
   $\{x | x \in R \text{ and } x > 3\}$

4. "Q is the set of all fractions in the form $\frac{a}{b}$ such that a and b are integers and b does not equal zero."

5. (a) counting (or natural) numbers (also whole numbers, integers, rational numbers, and real numbers)
   (b) integers (also rational numbers and real numbers)
   (c) negative integers (also rational numbers and real numbers)
   (d) irrational numbers (also real numbers)
   (e) same answer as (a). These are perfect squares.

\- - - - - - - - - - - -

SETS AND FLOWCHARTS

1. A flowchart is a diagram which consists of arrows and closed figures of different shapes. The arrows and figures are symbols which represent procedures to be followed in a data processing system. A flowchart is like a blueprint. It is a plan of action. We usually write instructions inside of or near each symbol.

   EXAMPLE: The purpose of the below flowchart is to show how a computer might process information contained in a set of punched IBM cards. Let each IBM card contain the name and sex of a person coded in holes punched in the card. Each card is read by a computer input device called a card reader. The information on the card is stored and examined in the central processing unit of the computer. The computer is programmed to select the names of male persons and print them on a computer output machine called the printer.

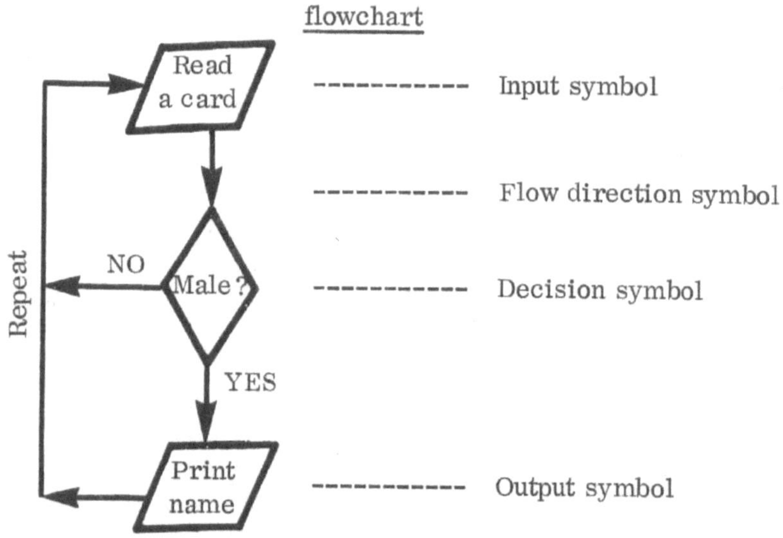

- - - - - - - - - - - - -

Note: While punched cards are no longer used in data processing applications today, the above example of decision making based on holes punched in a card is easy to understand.

2. Some fundamental concepts of sets can be applied to the Example that was given in Frame1.

EXAMPLES:

(a) All of the punched IBM cards form a universal set.
(b) The cards which are coded male form a subset of the universal set.
(c) The cards which are coded not male form a complementary set.

We can illustrate the above in set notation:

$U = \{\text{all punched IBM cards}\}$
$M = \{\text{punched IBM cards coded male}\}$
$\overline{M} = \{\text{punched IBM cards coded not male}\}$

$M \cup \overline{M} = U$ and $M \cap \overline{M} = \emptyset$.

We can illustrate this with a Venn diagram.

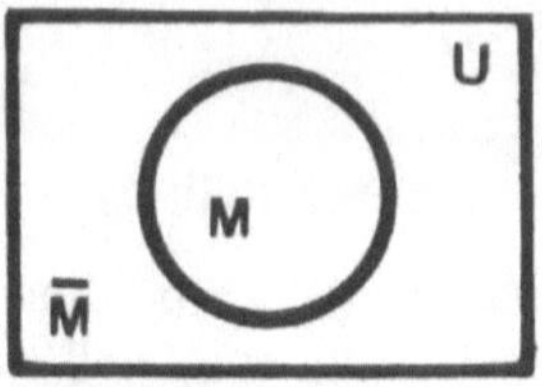

\- - - - - - - - - - - - -

Note: The reason why punched paper cards are no longer used to store data is the invention of a much less expensive magnetic and transistor storage media. This media can store the very same information that was stored on IBM punched cards for less cost and faster access to the data that is stored. For information on how numbers can be stored in modern computers, we cover this topic in *Computer Number Bases* and *An Introduction to Boolean Algebra and Switching Circuits*. Both texts by William R. Parks are part of our *Computer Mathematics Series*

3. The design of flowcharts can be made easier by using the principles of set theory.

EXAMPLE: Let us suppose that a life insurance company has 50,000 insured policy holders listed in its current file. The chief sales manager decides that his salesmen should contact some of the insured policy holders about buying more insurance. The problem is that the chief sales manager does not know which insured should be contacted.

An expert is called in and he proposes three possible solutions. The expert uses sets to explain his solutions. He tells the sales manager that the existing file of 50,000 insured policy holders could be used to make the decision. The expert says that his file of data cards is a set and that each card in the file is an element or member of the set.

EXPERT'S PRELIMINARY REMARKS: The file of data cards is called the universal set. The universal set is defined in set-builder notation.

$$U = \{d | d \text{ is an insured data card}\}$$

In this definition, each data card "d", would contain the following information: name of insured, sex, age, and amount of insurance. Three subsets of the above universal set are also defined.

$A = \{d | d \text{ is an insured data card of a male}\}$
$B = \{d | d \text{ is an insured data card of less than \$10,000 insurance}\}$
$C = \{d | d \text{ is an insured data card of under 40 years of age}\}$

Since A, B, and C are subsets of U, we can express this in set notation.

$$A \subseteq U, \quad B \subseteq U, \quad \text{and} \quad C \subseteq U.$$

- - - - - - - - - - - - -

Note: By applying set theory to data analysis we use a new approach for understanding flowcharting.

(Frame 3, contd.)

SOLUTIONS:

(a) In the first solution the manager is told to have his salesmen contact those insured who are male and who have under $10,000 insurance. In other words, the salesmen will contact only those insured who are members of both A and B. In set notation "A and B" is written as "$A \cap B$."

The flowchart of $A \cap B$ is as follows:

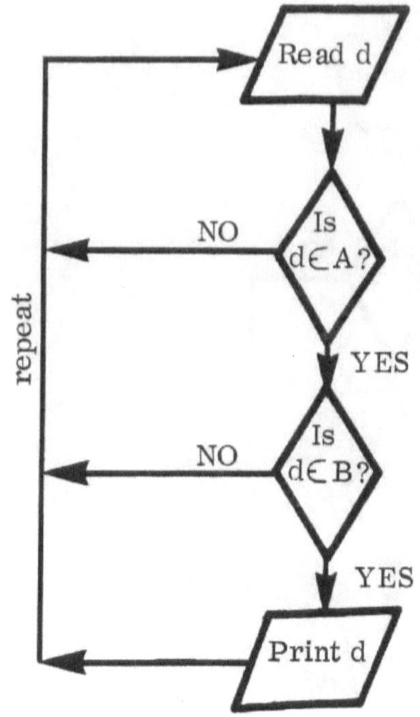

(Frame 3, contd.)

(b) In the second solution the manager is told to disregard sex and have his salesmen contact either an insured who had less than $10,000 insurance or an insured under 40 years of age. In other words, the salesmen are to contact members found in either B or C. In set notation, "B or C" is written as "$B \cup C$."

The flowchart of $B \cup C$ is as follows:

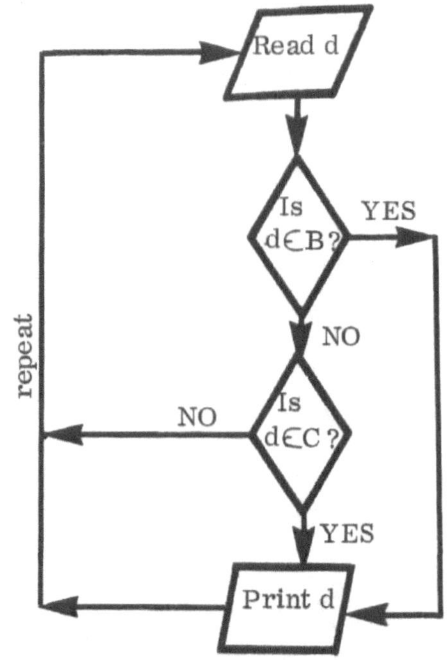

(Frame 3, contd.)

(c) In the third solution the manager is told to have his salesmen contact those insured who are male and who are either insured for less than \$10,000 or under 40 years of age. In other words, the salesmen are to contact members found in A <u>and</u> either B <u>or</u> C. In set notation, "in A <u>and</u> either B <u>or</u> C" is written as "in $A \cap (B \cup C)$."

The flowchart of $A \cap (B \cup C)$ is as follows:

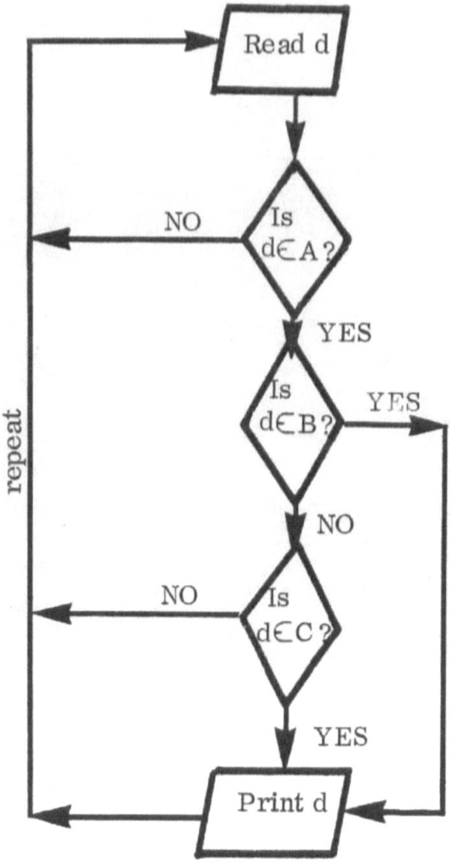

4. The below tables reveal what combinations of yes and no are needed before an insured data card is printed on the salesmen's contact list.

Each table refers to the corresponding flowchart in Frame 3.

(a) Table for $A \cap B$

$d \in A$?	$d \in B$?	$d \in (A \cap B)$?	
YES	YES	YES	(Print)
YES	NO	NO	(Do not print)
NO	YES	NO	(Do not print)
NO	NO	NO	(Do not print)

(b) Table for $A \cup B$

$d \in A$?	$d \in B$?	$d \in (A \cup B)$?	
YES	YES	YES	(Print)
YES	NO	YES	(Print)
NO	YES	YES	(Print)
NO	NO	NO	(Do not print)

(c) Table for $A \cap (B \cup C)$.

$d \in A$?	$d \in B$?	$d \in C$?	$d \in A \cap (B \cup C)$?	
YES	YES	YES	YES	(Print)
YES	YES	NO	YES	(Print)
YES	NO	YES	YES	(Print)
YES	NO	NO	NO	(Do not print)
NO	YES	YES	NO	(Do not print)
NO	YES	NO	NO	(Do not print)
NO	NO	YES	NO	(Do not print)
NO	NO	NO	NO	(Do not print)

- - - - - - - - - - - - -

5. EXERCISE

Answer all questions before checking the answer key in Frame 6.

1. A ________________________ is a diagram which consists of arrows and closed figures of different shapes.

2. The purpose of a flowchart might be to "blueprint" a plan that shows how a ________________________ might process punched IBM cards.

3. The problem of designing flowcharts can be made easier by applying the principles of ______________ ________________.

4. Suppose each data card in a data file contains the name, rank, age, and sex of an army officer. Design flowcharts for each of the given sets in (a), (b), and (c).

   Let the universal set U = $\{d | d$ is an officer's data card$\}$
   A = $\{d | d$ is a captain$\}$
   B = $\{d | d$ is under 35$\}$
   C = $\{d | d$ is female$\}$

   (a) $A \cap B$

   (b) $A \cup B$

   (c) $A \cap (B \cup C)$

5. Construct a table for each of the corresponding flowcharts in Question 4.

6. ANSWERS FOR EXERCISE

1. flowchart

2. computer

3. set theory

4. The answers are the same as the corresponding Solutions in Frame 3.

5. The answers are the same as the corresponding Tables in Frame 4.

\- - - - - - - - - - - - -

INTRODUCTION TO LOGIC

This section covers the fundamental concepts of logic used in human reasoning and especially in modern technological fields such as computer science, business data processing and computer applications. As stated at the beginning of the book, the material is presented in a simple and easy to understand format. We utilize a modified form of programmed instruction (PI).

Small bits of information are presented in each frame before advancing to the next frame. Exercises are listed after several frames, followed by an answer key.

We also give example problems with an immediate presentation of the solution right after the problem is stated. This immediate answer feedback approach to a problem is important when learning new concepts. The exercises serve the same purpose. Answers are given right after each set of exercises.

The study of logic goes hand and hand with the study of computer programming and basic concepts of flowcharting. The mathematics of set theory is also related to logic and is used to explain the basic principles of flowcharting for scientific and business data processing applications while observing basic rules of logic. Because of these interrelationships, we included this section in the computer math series.

INTRODUCTION TO LOGIC

1. In this study we will be concerned with symbolic logic. In symbolic logic letter symbols are used to represent declarative sentences.

   EXAMPLES:
   p: John is playing the violin.
   q: Mary is shutting the door.

   - - - - - - - - - - - - -

   Note: The letter "p" stands for the entire sentence, "John is playing the violin." The letter "q" stands for the entire sentence, "Mary is shutting the door."

2. Symbolic logic is concerned with statements which are either true or false, but not both.

   EXAMPLES:

(a) The White House is in Washington, D.C.	TRUE
(b) Chicago is in Texas.	FALSE
(c) What is your name?	NO TRUTH VALUE
(d) Please listen to me.	NO TRUTH VALUE

   - - - - - - - - - - - - -

   Note: It is true that the White House is located in Washington, D.C., therefore, we assign a truth value of "TRUE" to the sentence, "The White House is in Washington, D.C. The "TRUTH VALUE" of a sentence is either "true" or "false" but not both. "Truth value" is the "logical value" of a sentence which is either "true" or "false."

3. Two simple statements can be combined to form a compound statement. The following connectives are used between simple statements:

and : $\wedge$	conjunction
or : $\vee$	disjunction

EXAMPLES:

(a) John is playing the violin and Mary is shutting the door.
$p \wedge q$

(b) John is playing the violin or Mary is shutting the door.
$p \vee q$

- - - - - - - - - - - - - -

Note: The logical "and" in above Frame 3 looks like an upside down "V" shaped symbol. The term "conjunction" associated with this symbol refers to joining two statements with the "and" connective. The logical "or" looks like a "V" shaped symbol. The term disjunction associated with this symbol refers to joining two statements with the "or"connective.

4. The following connectives are also frequently used in symbolic logic:

not : $\sim$	negation	(refers to denial)
if-then : $\longrightarrow$	conditional	(refers to implication)

EXAMPLES:

(a) p : John is playing the violin.
$\sim p$: John is not playing the violin.

(b) p : John is playing the violin.
q : Mary is shutting the door.
$p \longrightarrow q$: If John is playing the violin, then Mary is shutting the door. (p is the hypothesis and q is the conclusion.)

- - - - - - - - - - - - - -

EXERCISE 1
Multiple choice questions:

1. Which symbols stand for sentences in symbolic logic.
(a) numerals (b) letters (c) words (d) none of these

2. Symbolic logic is concerned with statements that are either true or false.
(a) true (b) false

3. Which statement type uses the "and" connective.
(a) conditional (b) negation (c) conjunction (d) disjunction

4. Which statement type uses the "or" connective.
(a) conditional (b) negation (c) conjunction (d) disjunction

5. Which symbol stands for the if-then conditional.
(a) v (b) * (c) the arrow (d) none of these

6. Which is the negation of statement "p."
(a) not p (b) <P> (c) q (d) none of these

7. The compound statement "John is running and Mary is walking," can be represented by:
(a) $p \vee q$ (b) $p \wedge q$ (c) $p \rightarrow q$ (d) none of these

8. In symbolic logic a statement can be both true and false.
(a) true (b) false

9. In symbolic logic the statement $p \rightarrow q$ is called:
(a) a negation (b) a conjunction (c) a disjunction (d) a conditional

10. The negation symbol in symbolic logic is:
(a) $\rightarrow$ (b) $\wedge$ (c) $\sim$ (d) *

ANSWERS FOR EXERCISE 1

1. (b) letters

2. (a) true

3. (c) conjunction\

4. (d) disjunction

5. (c) the arrow

6. (a) not p

7. (b) $p \wedge q$

8. (b) false

9. (d) conditional

10. (c) ~

5. We can use p and q as previously defined and write compound statements which contain more than one connective.

EXAMPLES:

(a) $p \wedge (\sim q)$: John is playing the violin and Mary is not shutting the door.

(b) $(\sim q) \longrightarrow (\sim p)$: If Mary is not shutting the door, then John is not playing the violin.

(c) $[(p \longrightarrow q) \wedge p] \longrightarrow q$: If John plays the violin, then Mary shuts the door. John is playing the violin. Therefore, Mary is shutting the door.

- - - - - - - - - - - -

Note: In the above examples we connect statements p and q with the following connectives -- also called logical operators:

Negation (not): $\sim$

Conjunction (and): $\wedge$

Disjunction (or): $\vee$

Implication (if...then): $\rightarrow$

"And," (if…then) and "or" are called binary connectives because they join two sentences to make a compound statement. However, negation is a unary connective. It applies to a single statement and does not connect two statements.

- - - - - - - - - - - -

EXERCISE 2

Let p amd q stand for the same two sentences about John and Mary. Write the sentences in English for the following logical statements:

1. $(\sim p) \rightarrow (\sim q)$

2. $[\ (p \rightarrow q) \wedge (\sim p)\ ] \rightarrow (\sim q)$

3. $(\sim q) \rightarrow (\sim p)$

4. $p \wedge (\sim q)$

5. $\sim p$

6. q

- - - - - - - - - - - -

ANSWERS FOR EXERCISE 2

1. If John is not playing the violin, then Mary is not shutting the door.

2If John is playing the violin, then Mary is shutting the door. John is not playing the violin. Therefore, Mary is not shutting the door.

3. If Mary is not shutting the door, then John is not playing the violin.

4. John is playing the violin and Mary is not shutting the door.

5. John is not playing the violin.

6. Mary is shutting the door.

6. Truth tables are used in symbolic logic to define the meaning of connectives. In the following truth table:
p: I washed the windows.

(a) negation

p	~p
T	F
F	T

(I did not wash the windows.)

Note: In the above
However, when p is false then not p is true: (~p is true).

(b) conjunction

p	q	p ∧ q
T	T	T
T	F	F
F	T	F
F	F	F

(I washed the windows and cleaned the floor.)

RULE: The statement p ∧ q is true only if both p and q are true.

Note: A compound statement in logic is built from simple statements using the logical connectives. The truth value of p and q written as (p ∧ q) depends on the individual truth values of p and q.

(c) disjunction

p	q	p ∨ q
T	T	T
T	F	T
F	T	T
F	F	F

(I washed the windows or cleaned the floor.)

RULE: The statement p ∨ q is true if at least one of the two statements is true.

(d) conditional

p	q	p → q
T	T	T
T	F	F
F	T	T
F	F	T

(If I washed the windows, then I cleaned the floor.)

RULE: The statement p →q is false only when p is true and q is false.

EXERCISE 3
True or False questions:

1. The statement $p \wedge q$ is true only if both p and q are true.

2. If $\sim p$ is true then p is false.

3. If p is true and q is false then $p \vee q$ is false.

4. If p is true and q is false then the statement $p \rightarrow q$ is false

5. For two statements p and q there are four different possible combinations for conjunction, disjunction and conditionals.

ANSWERS FOR EXERCISE 3

1. True
2. True
3. False
4. True
5. True

7. Another connective that we use in symbolic logic is the biconditional. It is symbolized by $\leftrightarrow$. It connects two statements which are equivalent. Two statements are equivalent if they are either both true or both false.

EXAMPLE:

p	q	$p \leftrightarrow q$
T	T	T
T	F	F
F	T	F
F	F	T

(The game will be called off if and only if the players don't show up.)

- - - - - - - - - - - - -

Note: In logic the biconditional is a connective of two statements that taken together is true if and only if both statements have the same logical truth value. In the above truth table p "if and only" q is true when both p and q have the same truth value. When both p and q are true the biconditional is true. When both p and q are false the biconditional is also true. When both p and q have different truth values then such a biconditional is false.

- - - - - - - - - - - - -

EXERCISE 4:

1. Is the following a biconditional statement?
   If I do not eat then I will be hungry.

2. What is a biconditional statement?

3. When is a biconditional statement true?

4. Explain the difference between a conditional statement and a biconditional statement?

5. Give two examples of a biconditional statement?

ANSWERS FOR EXERCISE 4

1. No. This is an "if – then" conditional statement. The "if "part is a hypothesis and the then part is a conclusion. The "if – then" conditional statement is false only when the "if" part is true and the "then" part is false.

2. A biconditional statement has an operator connecting two statements. The operator is s logical connective designated by a double arrow ↔ and it means "if and only if."

3. A biconditional statement is true only when both statements have the same truth value, that is, when both statements are true or when both statements are false.

4. A biconditional statement is true only when both parts have the same truth value. A conditional statement is false only when the "if" part it true and the "then" part is false

5. I will get an "A" grade if and only if I get an "A" on the final exam.
A four sided figure is a square if and only if it has four sides of equal length.

8. By means of a truth table we can show that the statement $p \longrightarrow q$ is equivalent to its contrapositive $(\sim q) \longrightarrow (\sim p)$.

p	q	$p \longrightarrow q$	$\sim q$	$\sim p$	$(\sim q) \longrightarrow (\sim p)$
T	T	T	F	F	T
T	F	F	T	F	F
F	T	T	F	T	T
F	F	T	T	T	T

Note: The contrapositive of a conditional statement is formed by flipping the statements and negating each one. Thus we make the contrapositive of "If p then q" rendered as $p \rightarrow q$ by flipping it and negating each part to read: "If not q then not p." rendered as $\sim q \rightarrow \sim p$

The contrapositive of "If it is a square then it has four equal sides" is "If it does not have four equal sides then it is not a square." Both statements are equivalent. The conditional is equivalent to its contrapositive.

EXERCISE 5

1. What is the contrapositive of $m \rightarrow n$?

2. Explain what is the contrapositive of a conditional statement.

3. Give an example of a true conditional statement and state its contrapositive.

4. Are the two statements you gave in question 3 both true. If so then explain why?

5. True or False: The contrapositive of a true conditional statement is also true.

6. True or False: The contrapositive of a false conditional statement is also false.

ANSWERS FOR EXERCISE 5

1. $\sim n \rightarrow \sim m$

2. The contrapositive of a conditional statement is formed by flipping the statements and negating each one. Thus we make the contrapositive of "If p then q" rendered as $p \rightarrow q$ by flipping it and negating each part to read: "If not q then not p." rendered as $\sim q \rightarrow \sim p$

3. Conditional: If an object is an apple, then the object is a fruit.
Contrapositive: If the object is not a fruit, then the object is not an apple.

4. Both the conditional and the contrapositive statements in answer 3 are true. It is true that an apple is a fruit. It is true that if an object is not a fruit then it can not be an apple (because an apple is a fruit).

5. True.

6. True

9. An <u>argument</u> in symbolic logic consists of a set of given statements which are called <u>premises</u> from which is derived another statement called the <u>conclusion</u>. An argument is valid if and only if the conjunction of the premises implies the conclusion as a <u>tautology</u>. A tautology exists when an argument is always true for any truth values of its substatements.

Here is a simple example of an argument where the given two premises produce a conclusion:

Let p = I wash the windows
Let q = I clean the floor

First premise: $p \rightarrow q$ If I wash the windows then I clean the floor.

Second premise: p: I wash the windows

Conclusion: q: Therefore I clean the floor.

The argument can also be represented as follows:

Premise: If p then q
Premise: p
Therefore: q

This argument can also be represented as follows:

$[(p \rightarrow q) \wedge p] \rightarrow q$

[(p implies q) and p] implies q

EXERCISE 6

QUESTION 1: Test the validity of the following argument by constructing a truth table and show that it is a tautology.

p: John is a sophomore. q: John is taking biology.

ARGUMENTS:

(a) Given: $p \longrightarrow q$: If John is a sophomore, then he is taking biology.

Given: p: John is a sophomore.

Conclusion: q: John is taking biology.

ANSWER TO QUESTION NO. 1:

This argument is valid if and only if the statement $[(p \longrightarrow q) \wedge p] \longrightarrow q$ is a tautology.

p	q	$p \longrightarrow q$	$(p \longrightarrow q) \wedge p$	$[(p \longrightarrow q) \wedge p] \longrightarrow q$
T	T	T	T	T
T	F	F	F	T
F	T	T	F	T
F	F	T	F	T

The above valid argument is an illustration of the law of detachment (also called modus ponens). This law is a basic rule of inference in logic.

Note: In logic, the **Law of Detachment** says that the two statements are true, namely (1) If, then and (2). We can derive a third statement that is always true and therefore forms an all true tautology as indicated in the above last column of the truth table.

QUESTION 2: Test the validity of the following argument by constructing a truth table to show if it is a tautology.

Given: $p \longrightarrow q$: If John is a sophomore, then he is taking biology.

Given: q: John is taking biology.

Conclusion: p: John is a sophomoıe.

ANSWER TO QUESTION NO. 2:

This argument is not valid because the statement $[(p \rightarrow q) \wedge q] \rightarrow p$ is not a tautology.

p	q	$p \rightarrow q$	$(p \rightarrow q) \wedge q$	$[(p \rightarrow q) \wedge q] \rightarrow p$
T	T	T	T	T
T	F	F	F	T
F	T	T	T	F
F	F	T	F	T

Note: The last column of truth values has an F. Therefore, the statement is not a tautology and the argument is not valid.

\- - - - - - - - - - - -

CONCLUDING REMARKS:

Logic can be applied to the study of the natural sciences, mathematics, computer science, law and philosophy. Its origin dates back to the time of the ancient Greek philosophers, especially noteworthy is Aristotle. The study of logic has developed over the centuries into a formal course with applications in many areas of knowledge.

COMPUTER NUMBER BASES

Base Ten Numbers

1. The decimal number system is a system of counting that is based on ten symbols called decimal digits. They are 0, 1, 2, 3, 4, 5, 6, 7, 8, and 9. Another name for decimal number system is base ten.

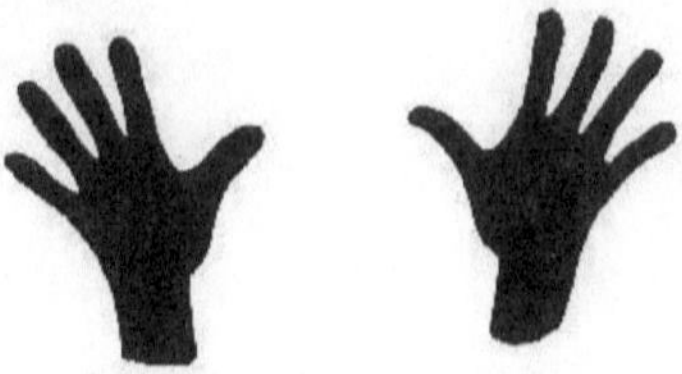

\- - - - - - - - - - - - -

2. The method used to represent decimal numbers greater than 9 is based on the place value principle.

EXAMPLE: The decimal number 6,575 contains two fives. However, each five is located in a different place. The place value is different for each five.

hundreds ones — different place values

6, 5 7 5

\- - - - - - - - - - - - -

3. The value of each place in a decimal number is ten times the place value of the position to the right of it.

EXAMPLE: The number 6,575 has four places. The place value of each position is as follows:

thousands hundreds tens ones
6, 5 7 5

It becomes more evident that each position increases ten times in place value when we use decimal numerals instead of words.

1000 100 10 1
6, 5 7 5

---

4. A decimal number can be expressed in what is called expanded notation.

EXAMPLE:

6,575 = 6 thousands + 5 hundreds + 7 tens + 5 ones
= 6000 + 500 + 70 + 5

---

5. The expanded notation can be written in other ways.

EXAMPLE:

6,575 = 6000 + 500 + 70 + 5
= (6 x 1000) + (5 x 100) + (7 x 10) + (5 x 1)
= (6 x 10 x 10 x 10) + (5 x 10 x 10) + (7 x 10) + (5 x 1)

---

6. An <u>exponent</u> is a number that tells us how many times another number is multiplied. Expanded notation with exponents can be used in place of the longer expanded notation that lists all the factors of ten.

EXAMPLE:

$$6,575 = (6 \times 10 \times 10 \times 10) + (5 \times 10 \times 10) + (7 \times 10) + (5 \times 1)$$
$$= (6 \times 10^3) + (5 \times 10^2) + (7 \times 10^1) + (5 \times 10^0)$$

NOTE: $1000 = 10 \times 10 \times 10 = 10^3$

$100 = 10 \times 10 = 10^2$

$10 = 10^1$

$1 = 10^0$

- - - - - - - - - - - - -

7. For decimal numbers that have more than four places, we can observe the same pattern of place values. The place value of each place is ten times the place value of the position to the right of it.

EXAMPLE:

millions	hundred thousands	ten thousands	thousands	hundreds	tens	ones
6,	5	4	8,	3	4	2

1,000,000	100,000	10,000	1000	100	10	1
6,	5	4	8,	3	4	2

10^6	10^5	10^4	10^3	10^2	10^1	10^0
6,	5	4	8,	3	4	2

- - - - - - - - - - - - -

8. EXERCISE 1

Answer all questions before checking the answer key in Frame 9.

1. The symbols used in the base ten system of counting are called ______________________ digits.

2. Each position in a decimal number has a defined ________ value.

3. The value of each position in a decimal number is _______ times the place value of the position to the right of it.

4. Using words, name the place value of each position in the following decimal number:

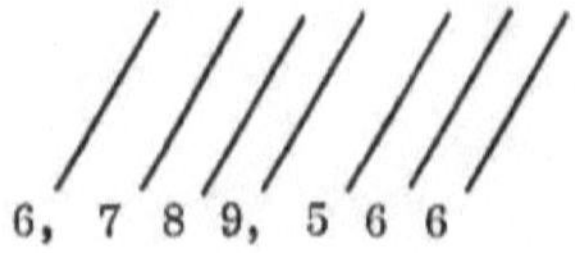

5. Express the following numbers in expanded notation. Refer to Frames 4, 5, and 6 and practice with as many expanded notation formats as were given:

   (a) 57 =

   (b) 698 =

   (c) 2,526 =

   (d) 3,456,789 =

6. The decimal number system is based on powers of _______.

9. ANSWERS FOR EXERCISE 1

1. decimal

2. place

3. ten

4.

millions	hundred thousands	ten thousands	thousands	hundreds	tens	ones
6,	7	8	9,	5	6	6

5. (a) 57 = 5 tens + 7 ones
$= 50 + 7$
$= (5 \times 10) + (7 \times 1)$
$= (5 \times 10^1) + (7 \times 10^0)$

(b) 698 = 6 hundreds + 9 tens + 8 ones
$= 600 + 90 + 8$
$= (6 \times 100) + (9 \times 10) + (8 \times 1)$
$= (6 \times 10^2) + (9 \times 10^1) + (8 \times 10^0)$

(c) 2,526 = 2 thousands + 5 hundreds + 2 tens + 6 ones
$= 2000 + 500 + 20 + 6$
$= (2 \times 1000) + (5 \times 100) + (2 \times 10) + (6 \times 1)$
$= (2 \times 10^3) + (5 \times 10^2) + (2 \times 10^1) + (6 \times 10^0)$

(d)

1,000,000	100,000	10,000	1000	100	10	1
3,	4	5	6,	7	8	9

10^6	10^5	10^4	10^3	10^2	10^1	10^0
3,	4	5	6,	7	8	9

6. ten

Base Two Numbers

10. The binary number system is a system of counting that is based on two symbols called binary digits (they are 0 and 1). Another name for binary number system is base two.

0	1
OFF	ON

- - - - - - - - - - - - -

11. The problem of representing numbers bigger than 1 in base two is solved by using the place value principle. However, the binary place value principle is based on powers of 2 (in the decimal system, the place value is based on powers of 10.)

EXAMPLE: The number $1001_{\text{base two}}$ contains two 1's. Each 1 is located in a different place. The place value is different for each 1.

eights   ones

$1\ 0\ 0\ 1_{\text{base two}}$

- - - - - - - - - - - - -

12. The value of each place in a binary number is two times the place value of the position to the right of it.

EXAMPLE: The number $1001_{\text{base two}}$ has four places. The place value of each position is as follows:

eights fours twos ones

$1\ 0\ 0\ 1_{\text{base two}}$

It becomes more evident that each position increases two times in place value when we use numerals in place of words.

8 4 2 1

$1\ 0\ 0\ 1_{\text{base two}}$

- - - - - - - - - - - - -

13. A binary number can be converted to base ten by <u>expanded notation</u>.

EXAMPLE:

$$\begin{array}{l} \;\;8\;\;4\;\;2\;\;1 \\ 1\;\;0\;\;0\;\;1_{\text{base two}} = (1 \times 8) + (0 \times 4) + (0 \times 2) + (1 \times 1) \\ \qquad\qquad\qquad = 8 + 0 + 0 + 1 \\ \qquad\qquad\qquad = 9_{\text{base ten}} \end{array}$$

- - - - - - - - - - - - -

14. The expanded notation method of converting to base ten can be written in another way.

EXAMPLE:

$$\begin{array}{l} \;\;8\;\;4\;\;2\;\;1 \\ 1\;\;0\;\;0\;\;1_{\text{base two}} = (1 \times 8) + (0 \times 4) + (0 \times 2) + (1 \times 1) \\ \qquad\qquad\qquad = (1 \times 2^3) + (0 \times 2^2) + (0 \times 2^1) + (1 \times 2^0) \end{array}$$

NOTE: $8 = 2 \times 2 \times 2 = 2^3$, $4 = 2 \times 2 = 2^2$, $2 = 2^1$, and $1 = 2^0$.

- - - - - - - - - - - - -

15. For binary numbers that have more than four places, we can observe the same pattern of place values. The value of each place is <u>two times</u> the place value of the position to the right of it.

EXAMPLE:

$$\begin{array}{l} \qquad\qquad\; 2^6\;2^5\;2^4\;2^3\;2^2\;2^1\;2^0 \qquad 64\;32\;16\;8\;4\;2\;1 \\ 1010111_2 = 1\;0\;1\;0\;1\;1\;1_2 = 1\;0\;1\;0\;1\;1\;1_2 \\ \qquad\qquad = 64 + 0 + 16 + 0 + 4 + 2 + 1 \\ \qquad\qquad = 87_{10} \end{array}$$

- - - - - - - - - - - - -

16. EXERCISE 2

Answer all questions before checking the answer key in Frame 17.

1. The binary number system is a system of counting based on two symbols (0 and 1) which are called ______________ digits.

2. Each position in a binary number has a defined __________ value.

3. The place value of each position in a binary number is _____ times the place value of the position to the right of it.

4. Name the place value of each position in the following binary number. Use decimal numerals.

$$1\ 0\ 1\ 1\ 1\ 0\ 1_2$$

5. Convert the following numbers to base ten by the expanded notation method. Refer to Frames 12 through 15 and practice with as many expanded notation formats as were given:

   (a) 1111_2

   (b) 10010_2

   (c) 1010101_2

   (d) 1111101_2

6. The binary number system is based on powers of ________.

\- - - - - - - - - - - -

17. ANSWERS FOR EXERCISE 2

1. binary

2. place

3. two

4. 64 32 16 8 4 2 1

   $1\ 0\ 1\ 1\ 1\ 0\ 1_2$

5. 8 4 2 1

   (a) $1\ 1\ 1\ 1_2 = 8 + 4 + 2 + 1 = 15$

   (or) $1111_2 = (1 \times 2^3) + (1 \times 2^2) + (1 \times 2^1) + (1 \times 2^0)$
   $= 8 + 4 + 2 + 1 = 15$

   16 8 4 2 1

   (b) $1\ 0\ 0\ 1\ 0_2 = 16 + 0 + 0 + 2 + 0 = 18$

   (or) $10010_2 = (1 \times 2^4) + (0 \times 2^3) + (0 \times 2^2) + (1 \times 2^1) + (0 \times 2^0)$
   $= 16 + 0 + 0 + 2 + 0 = 18$

   64 32 16 8 4 2 1

   (c) $1\ 0\ 1\ 0\ 1\ 0\ 1_2 = 64 + 0 + 16 + 0 + 4 + 0 + 1 = 85$

   (or) $1010101_2 = (1 \times 2^6) + (0 \times 2^5) + (1 \times 2^4) + (0 \times 2^3) + (1 \times 2^2) + (0 \times 2^1) + (1 \times 2^0)$
   $= 64 + 0 + 16 + 0 + 4 + 0 + 1 = 85$

   64 32 16 8 4 2 1

   (d) $1\ 1\ 1\ 1\ 1\ 0\ 1_2 = 64 + 32 + 16 + 8 + 4 + 0 + 1 = 125$

   (or) $1111101_2 = (1 \times 2^6) + (1 \times 2^5) + (1 \times 2^4) + (1 \times 2^3) + (1 \times 2^2) + (0 \times 2^1) + (1 \times 2^0)$
   $= 64 + 32 + 16 + 8 + 4 + 0 + 1 = 125$

6. two

- - - - - - - - - - - - -

Base Eight Numbers

18. The octal number system is a system of counting that is based on eight symbols called octal digits. They are 0, 1, 2, 3, 4, 5, 6, 7. Another name for octal number system is base eight.

base two to base eight

- - - - - - - - - - - -

19. The method used to represent octal numbers greater than 7 is based on the place value principle. The octal place value principle is based on powers of eight.

EXAMPLE: The number $343_{\text{base eight}}$ contains two 3's.

Each 3 is in a different place. The place value is different for each 3.

sixty-fours eights ones

$343_{\text{base eight}}$

- - - - - - - - - - - -

20. The value of each place in an octal number is eight times the place value of the position to the right of it.

EXAMPLE:

64 8 1

$343_{\text{base eight}}$

- - - - - - - - - - - -

24. An easy method exists for converting from base two to base eight. In this method we group the binary digits in threes.

EXAMPLE: To convert 11100110_2 to base eight, group the binary digits in threes and write the place values of each group as follows:

$$
\begin{array}{lccccccccc}
 & & 2 & 1 & 4 & 2 & 1 & 4 & 2 & 1 \\
11100110_2 & = & 1 & 1 & 1 & 0 & 0 & 1 & 1 & 0 \\
 & = & & 3 & & 4 & & & 6 &
\end{array}
$$

Therefore $11100110_2 = 346_8$

NOTE: We begin grouping on the right and move to the left.

\- - - - - - - - - - - - -

25. An inverse operation exists for converting from base eight back to base two. The principle is the same as in the last frame.

EXAMPLE: To convert 26473_8 to base two, take each octal digit and convert it to a three digit binary number as follows:

$$
\begin{array}{lcccccc}
 & & 2 & 6 & 4 & 7 & 3 \\
26473_8 & = & 010 & 110 & 100 & 111 & 011
\end{array}
$$

Therefore $26473_8 = 10110100111011_2$

\- - - - - - - - - - - - -

26. EXERCISE 3

Answer all questions before checking the answer key in Frame 27.

1. The octal number system is a system of counting based on eight symbols which are called ________________ digits.

2. Each position in an octal number has a defined __________ value.

3. The place value of each position in an octal number is _____ times the place value of the position to the right of it.

4. Name the place value of each position in the following octal number. Use decimal numerals.

$$6\ 4\ 7\ 2_8$$

5. Convert the following numbers to base ten by the expanded notation method. Refer to Frames 21 through 23 and practice with as many expanded notation formats as were given:

   (a) 24_8

   (b) 522_8

   (c) 3627_8

6. Convert the following numbers to base eight:

   (a) 1011_2

   (b) 10111010_2

   (c) 10110010100101_2

(Frame 26, contd.)

7. Convert the following numbers to base two:

   (a) 5_8

   (b) 43_8

   (c) 6325_8

   (d) 637214_8

8. The octal number system is based on powers of ________.

- - - - - - - - - - - - -

27. ANSWERS FOR EXERCISE 3

1. octal

2. place

3. eight

4.
   512 64 8 1
   $6\ 4\ 7\ 2_8$

5. (a) 20_{10} (b) 338_{10} (c) 1943_{10}

6. (a) 13_8 (b) 272_8 (c) 26245_8

7. (a) 101_2 (b) 100011_2 (c) 110011010101_2

   (d) 110011111010001100_2

8. eight

- - - - - - - - - - - - -

Base Sixteen Numbers

28. The hexadecimal number system is a system of counting that is based on sixteen symbols called hexadecimal digits. They are: 0, 1, 2, 3, 4, 5, 6, 7, 8, 9, A, B, C, D, E, F. Another name for hexadecimal number system is base sixteen.

Hexadecimal Digit	Decimal Value
0	0
1	1
2	2
3	3
4	4
5	5
6	6
7	7
8	8
9	9
A	10
B	11
C	12
D	13
E	14
F	15

- - - - - - - - - - - -

29. The method used to represent hexadecimal numbers greater than F is based on the place value principle. The hexadecimal place value principle is based on powers of sixteen.

EXAMPLE: The number $2A2_{\text{base sixteen}}$ contains two 2's. Each 2 is in a different place. The place value is different for each 2.

two hundred fifty-sixes
sixteens
ones

$2\ A\ 2_{\text{base sixteen}}$

- - - - - - - - - - - - -

30. The value of each place in a hexadecimal number is sixteen times the place value of the position to the right of it.

EXAMPLE:

256 16 1

$2\ A\ 2_{\text{base sixteen}}$

- - - - - - - - - - - - -

31. A hexadecimal number can be converted to base ten by expanded notation.

EXAMPLE:

256 16 1

$$
\begin{aligned}
2\ A\ 2_{\text{base sixteen}} &= (2 \times 256) + (10 \times 16) + (2 \times 1) \\
&= 512 + 160 + 2 \\
&= 674_{\text{base ten}}
\end{aligned}
$$

- - - - - - - - - - - - -

32. The expanded notation method of converting to base ten can be written in another way.

EXAMPLE:

256 16 1

$2\ A\ 2_{\text{base sixteen}} = (2 \times 256) + (10 \times 16) + (2 \times 1)$

$= (2 \times 16^2) + (10 \times 16^1) + (2 \times 16^0)$

NOTE: $256 = 16 \times 16 = 16^2$, $16 = 16^1$, and $1 = 16^0$

- - - - - - - - - - - -

33. An easy method exists for converting from base two to base sixteen. In this method we group the binary digits in fours.

EXAMPLE: To convert 1011100110_2 to base sixteen, group the binary digits in fours and write the place values of each group as follows:

2 1   8 4 2 1   8 4 2 1

$1011100110_2 = 1\ 0 \quad 1\ 1\ 1\ 0 \quad 0\ 1\ 1\ 0$

$= 2 \quad E \quad 6$

Therefore $1011100110_2 = 2E6_{16}$

NOTE: We begin grouping on the right and move to the left.

- - - - - - - - - - - -

34. An inverse operation exists for converting from base sixteen back to base two. The principle is the same as in the last frame.

EXAMPLE: To convert $3A8E_{16}$ to base two, take each hexadecimal digit and convert it to a four digit binary number as follows:

3   A   8   E

$3A8E_{16} = 0011 \quad 1010 \quad 1000 \quad 1110$

Therefore $3A8E_{16} = 11101010001110_2$

- - - - - - - - - - - -

35. EXERCISE 4

Answer all questions before checking the answer key in Frame 36.

1. The place value of each position in a hexadecimal number is ______________ times the place value of the position to the right of it.

2. The hexadecimal number system is based on powers of _____.

3. Change to base ten.

   (a) $7B_{16}$ =

   (b) $2FC_{16}$ =

4. Change to base two.

   (a) $7B_{16}$ =

   (b) $2FC_{16}$ =

5. Change to base sixteen.

   (a) 1111111_2 =

   (b) 1011111000_2 =

- - - - - - - - - - - - -

36. ANSWERS FOR EXERCISE 4

1. sixteen

2. sixteen

3. (a) 123_{10} (b) 764_{10}

4. (a) 1111011_2 (b) 1011111100_2

5. (a) $7F_{16}$ (b) $2F8_{16}$

BOOLEN ALGEBRA AND SWITCHING CIRCUITS

Boolean algebra is named after George Boole (1815-1864), a mathematician and philosopher who wrote, *The Laws of Thought*, which is the basis for Boolean algebra and logic used to design modern day computers.

George Boole (1815-1864)

In elementary algebra, 1 + 1 equals 2. In Boolean algebra, 1 + 1 equals 1.

Elementary algebra includes the study of variables and basic operations of arithmetic such as addition and multiplication. The variables can stand for a variety numbers such as the integers 1, 2, 3 ... or decimals numbers such as .5, 1.5, 2.7, etc. The basic rules of arithmetic apply in elementary algebra.

However, when we study Boolean algebra and switching circuits the values of variables are limited to two values denoted by the binary digits "1" and "0."

"1" can be represented in the physical world by a closed switch which will turn on a light and "0" can be represented by an open switch which will turn off a light.

The smallest unit in a digital computer is the binary digit which has only two possible states: 0 and 1. These two bits are the smallest units of information in a computer which can be stored or transmitted as high or low voltages representing 1 or 0.

Boolean Algebra is directly related to the study of propositional logic where truth values are studied. The symbol "1" in Boolean algebra is considered "true" in propositional logic and the meaning of "0" is "false" in logic. For this reason, truth tables used in logic can be used to to find the values of Boolean expressions.

Not all rules of elementary algebra work in Boolean algebra. Boolean algebra is not isomorphic (similar) to elementary algebra. However, Boolean algebra is isomorphic to logic.

In Boolean algebra and switching circuits there are three main operations that are also found in the study of propositional logic. In Boolean algebra the plus sign "+" means "OR" (disjunction in logic) and the times sign "**.** " means "AND" (conjunction in logic) & the prime mark or tilde "~" means "NOT" (negation in logic).

Conjunction in logic is denoted by an up arrow "∧" and as stated above stands for "AND." Disjunction in logic is denoted by a down arrow "∨" and stands for "OR." Negation, a unary operation, is often denoted by "~" (tilde symbol) and it stands for "NOT" in logic.

Logic symbols will be used mainly in Volume 3, *Introduction to Logic,* which is part of this *Computer Mathematics* Series. Logic symbols are directly related to corresponding symbols in Boolean algebra.

Boolean algebra and logic have different symbols to represent the same concepts. Some books on Boolean algebra use corresponding logic symbols. This might confuse beginning students.

In this "Computer Mathematics Series" new information is presented in short sections with immediate testing. This form of personalized instruction is often used in on-line Internet based courses for distance learning. A small amount of information is presented in each section before advancing to the next section. Exercises are listed after several sections followed by an answer key.

BOOLEAN ALGEBRA AND SWITCHING CIRCUITS

1. In this study we will be concerned with switching circuits. A switch is either a mechanical or electrical device which can be either open or closed.

ILLUSTRATIONS:

open switch

closed switch

----

2. Variables are used to represent switches because a switch's condition varies. It can be in one of two states, "0" corresponding to open and "1" corresponding to closed.

EXAMPLES:

(a) —— x —— (Switch x may be open or closed.)

If the above switch is open, then x = 0.
If the above switch is closed, then x = 1.

(b) —— y —— (Switch y may be either 0 or 1.)

"1" represents in the physical world a closed switch which can turn on a light and "0" represents an open switch which can turn off a light. In a computer they can be stored or transmitted as high or low voltages representing 1 or 0.

1 and 0 are called binary digits. Larger decimal numbers are stored as a series of binary digits. This topic is covered in *Computer Number Bases* in the four volume series.

----

3. Two switches x, y may be connected together either in series or in parallel.

EXAMPLES:

(a) 

(x and y are two switches in series)

(b) 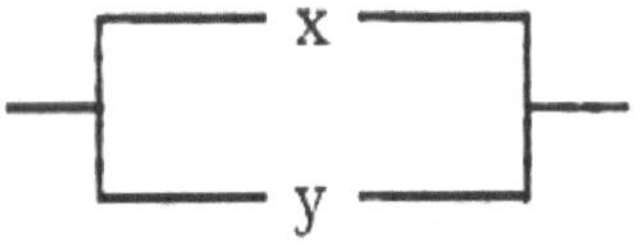

(x and y are two switches in parallel.)

4. There are four possible combinations of 0 and 1 for two switches x, y that can be configured together in series or in parallel as shown in the following table.

x	y
1 (closed)	1 (closed)
1 (closed)	0 (open)
0 (open)	1 (closed)
0 (open)	0 (open)

5. EXERCISE 1

1. A switch can be open or closed.

(a) True (b) False

2. Which of the following stand for switching states in Boolean algebra:

(a) 2 and 4 (b) 1 and 0 (c) 5 and 15

3. Variables (i.e. letter symbols) are used to represent switches because a switch can be in one of two states.

(a) True (b) False

4. How many different possible combinations are there for two switches x and y.

(a) One (b) Two (c) Three (d) Four

5. Two switches x and y may be connected:
(a) Only in series (b) Only in parallel (c) In series or in parallel

6. ANSWERS FOR EXERCISE 1

1. (a) True

2. (b) 1 and 0

3. (a) True

4. (d) Four: (1, 1), (1,0), (0, 1), (0,0)

5. (c) In series or in parallel

\- - - - - - - - - - - - -

7. We denote two switches connected in parallel by the plus sign +.

EXAMPLE:

The expression "x + y" represents

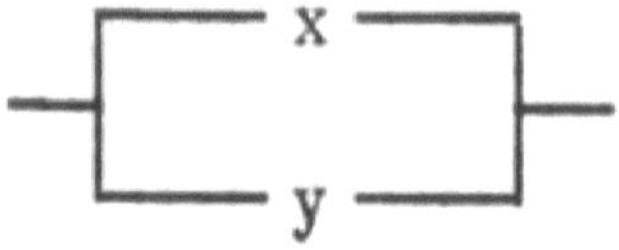

----

8. The table on the next page representing x + y indicates when the parallel circuit is either open or closed. There are four possible combinations of 0 and 1 for two switches x and y. In this table we list the four possible combinations in the first two columns in the same order as was presented in Section 4.

x	y	x + y	
1	1	1	(closed)
1	0	1	(closed)
0	1	1	(closed)
0	0	0	(open)

In the first horizontal line below the heading , switch x is one. It is closed. Switch y is also one and it is closed. Therefore, in the third column x + y must also be closed and equal to one. Therefore,

an electrical current can pass through a parallel circuit x + y if both switches are closed. Also, the current flows if just one of the switches is closed. Therefore, in Boolean algebra: 1 + 1 = 1; 1 + 0 = 1 and 0 + 1 = 1.

RULE: The parallel circuit x + y in the above table "equals"one, if at least one of the two switches "equals" one.

The circuit at the top of the next page and the accompanying equation represents the last case in the above table. The equation, 0 + 0 = 0, means "an open switch in parallel with an open switch is open." No current may pass through.

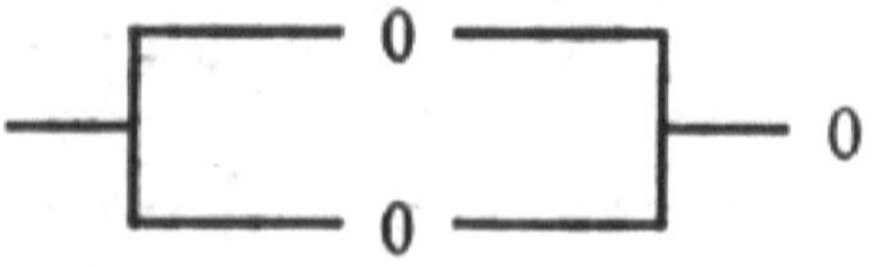

0 + 0 = 0

---

9. We denote two switches connected in series by a times sign " · ".

EXAMPLE:

The expression "x · y" represents

—— x —— y ——

The times sign means that the two switches x and y are connected in series.

For electrical circuits this means that an electrical current is provided with only one path.

10. The following table for x · y indicates under which circumstances a <u>series circuit</u> is either open or closed.

x	y	x · y	(series circuit)
1	1	1	(closed)
1	0	0	(open)
0	1	0	(open)
0	0	0	(open)

The first two columns in this table are the same as in the previous table in Section 6.

However, in the third column we see a different set of binary digits. The only time a one or closed circuit condition occurs is when switches x and y are both one.

Rule: The series circuit x · y "equals" one only if both x and y are one.

—— 1 —— 1 —— 1 $\quad$ $1 \cdot 1 = 1$

This circuit and the above equation represents the first case in the above series circuit table. The equation $1 \cdot 1 = 1$ means "a closed switch in series with a closed switch is closed. This case will allow current to flow and turn on a light bulb .

If we assume it's a complete circuit with electricity flowing through the wire, the light bulb will be on.

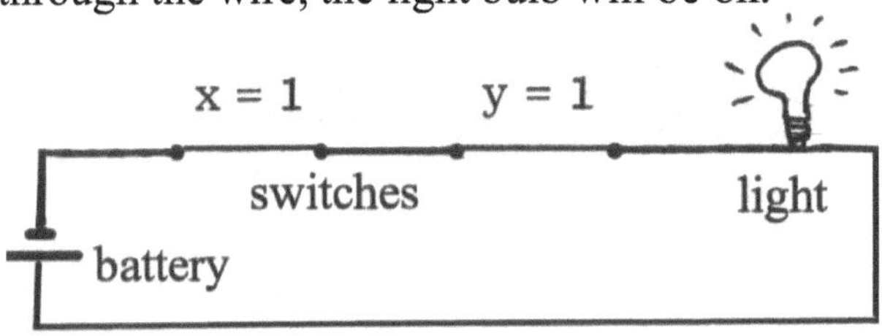

11. EXERCISE 2

1. We denote two switches in parallel by a

(a) + sign (b) - sign (c) x sign (d) None of these

2. A parallel circuit x + y always equals one.

(a) True (b) False

3. A series circuit equals one when all switches equal one.

(a) True (b) False

4. If switch x is on and switch y is off in a parallel circuit then x + y = 1.

(a) True (b) False

12. ANSWERS FOR EXERCISE 2

1. (a) + sign

2. (b) False

3. (c) True

4. (a) True

13. Switches in a circuit which are always in opposite states are called complementary switches. When the first switch is closed, then the second switch will be open.

In the below table switch x-prime represents the opposite of x. The prime mark is used to designate a switch in opposite state. The tilde mark "~" that is often used in logic can be interpreted to mean the negation of x. In this sense it is a unary operation operating on one switch. In previous examples we learned that the plus and times symbols are binary operations operating on two switches.

EXAMPLE:

x	x'
1	0
0	1

In the first line of the above table, x is 1 and x-prime is 0. Switch x is closed and switch x-prime its complement is open. In the next line switch x is 0 (or open) and switch x-prime (its complement) is 1 (or closed).

Let's examine the following two Boolean equations. The first equation is read, "x plus x-prime equals 1." This means that an electrical current will pass through a parallel circuit consisting of two complementary switches.

$$x + x' = 1$$
$$x \cdot x' = 0$$

The second equation, "x times x-prime equals 0." This means that an electrical circuit can never pass through a series circuit consisting of two complementary switches.

\- - - - - - - - - - - -

14. Series and parallel circuits may be combined to form more complex circuits.

EXAMPLES:

(a) Series-parallel circuit:

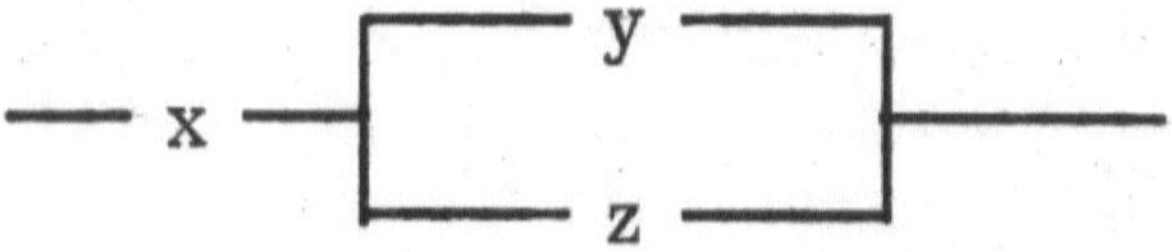

In the above circuit.switches y and z are in parallel and can be represented by (y + z). This parallel circuit is in series with x.

The entire circuit is represented by the expression:

$x \cdot (y + z)$

(b) The circuit $x \cdot y + z$ is represented by a different configuration.as follows:

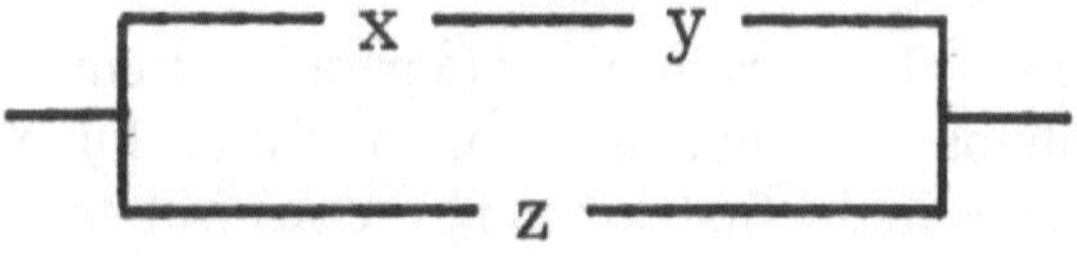

We interpret the expression $x \cdot y + z$ by deciding that x times y has precedence over y + z.

RULE: When more than one binary operator is used in a Boolean expression to represent a circuit, "times" is evaluated first and "plus" next. In logic, the unary operator NOT is evaluated first, then the binary operator AND which is the same as "times" in Boolean algebra. Finally OR is evaluated which is the same as a plus in Boolean algebra..

\- - - - - - - - - - - - -

15. The next table shows eight possible combinations when three switches are configured in the following series-parallel circuit:

$$x \cdot (y + z)$$

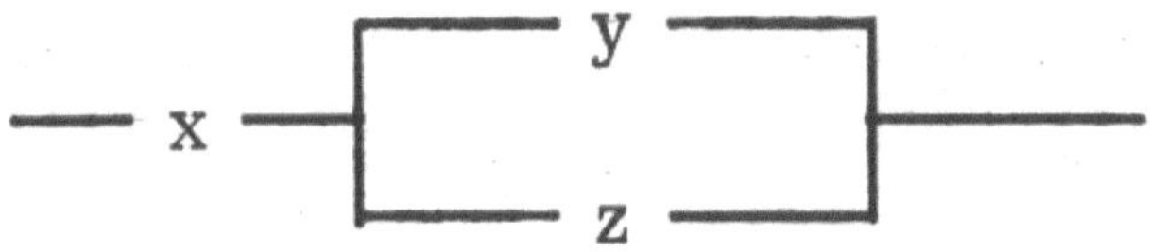

x	y	z	y + z	x · (y + z)
0	0	0	0	0
0	0	1	1	0
0	1	0	1	0
0	1	1	1	0
1	0	0	0	0
1	0	1	1	1
1	1	0	1	1
1	1	1	1	1

Here the corresponding set of values for y + z (a parallel circuit) are shown in the fourth column. The final results in the fifth column are calculated using the values in the first column in series with the values of the fourth column.

This table indicates clearly the exact order of “on” and “off” settings that will permit a current to flow through this three switch circuit. The combinations in the last three rows permit a current to pass through this particular configuration of switches. x has to be 1 in each case and at least one of either y or z switches must be 1.

In the above table, we placed zeros in the first row. This is arbitrary. In future tables the digits will be shown in reverse order. Three ones will be placed in the first row.

16. EXERCISES 3

1. Complementary switches in a circuit are always in opposite states.

(a) True (b) False

2. Series and parallel circuits may not be combined.

(a) True (b) False

3. How many combinations of 0 and 1 are possible with three switches?

(a) 4 (b) 6 (c) 8 (d) 10

4. In the circuit $x \cdot (y + z)$, if $x = 1$, $y = 0$ and $z = 1$, the circuit will be

5. In the same circuit $x \cdot (y + z)$, if $x = 0$, $y = 1$ and $z = 1$, the circuit will be:

(a) "0" and will not pass current (b) "1" and will pass current

6. In the circuit $x \cdot y + z$, which binary operator has precedence and is evaluated first:

(a) $x \cdot y$ is evaluated first (b) $y + z$ is evaluated first

7. The circuit $x \cdot y + z$ is represented by the following diagram

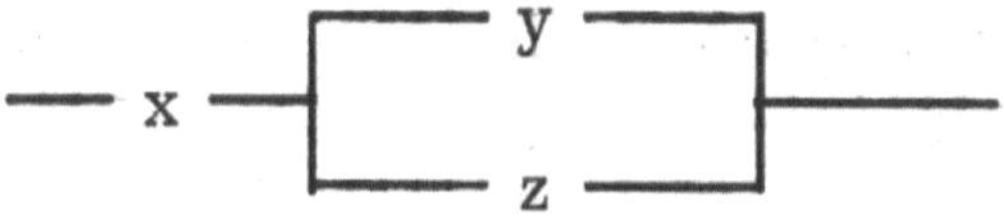

(a) True (b) False

17. ANSWERS FOR EXERCISE 3

1. (a) True

2. (b) False

3. (c) 8

4. (b) "1" and will pass current

5. (a) "0" and will not pass current

6. (a) x · y is evaluated first

7. (b) False

18. One purpose of Boolean algebra tables is to show that circuits are equivalent. If two columns representing circuits in a given table are identical, then we can say that their circuits are equivalent.

EXAMPLES:

(a) The following table shows x • y + x is equivalent to x because the 0 and 1 entries are identical for each column. Electrical current will flow in both circuits as indicated in the first two rows.

x	y	x · y	x · y + x	x
1	1	1	1	1
1	0	0	1	1
0	1	0	0	0
0	0	0	0	0

(b) The next table shows $x \cdot (y + z) = x \cdot y + x \cdot z$

The equivalence of these two circuits is called the first distributive property. This property is also used in elementary algebra (x is distributed over y and z).

x	y	z	y + z	x · (y + z)	x · y	x · z	x · y + x · z
1	1	1	1	1	1	1	1
1	1	0	1	1	1	0	1
1	0	1	1	1	0	1	1
1	0	0	0	0	0	0	0
0	1	1	1	0	0	0	0
0	1	0	1	0	0	0	0
0	0	1	1	0	0	0	0
0	0	0	0	0	0	0	0

The fifth column of the binary digits ones and zeros ("on" and "off") is identical in sequence to the binary digits in the last column. Therefore, the two circuits are equivalent.

The two distributive property expressions for these two circuits are shown in the following circuit diagrams. The two circuit diagrams appear to be different, however they are equivalent as shown in the previous table.

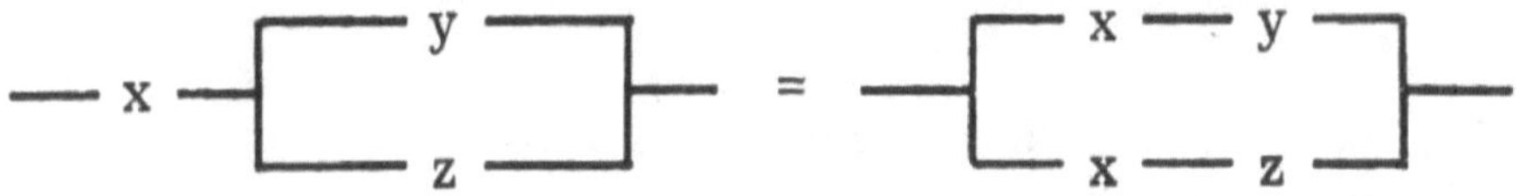

$$x \cdot (y + z) \quad = \quad x \cdot y + x \cdot z$$

----

19. The two sides of the below circuit equation appear to be different, however they are equivalent because each side produces the same table values. The equivalence of these two circuits is called the second distributive property.

The second distributive property of Boolean algebra is not allowed in elementary algebra. "x +" is distributed over y and z in Boolean algebra.

$$x + (y \cdot z) = (x + y) \cdot (x + z)$$

x	y	z	y · z	x + (y · z)	x + y	x + z	(x + y) · (x + z)
1	1	1	1	1	1	1	1
1	1	0	0	1	1	1	1
1	0	1	0	1	1	1	1
1	0	0	0	1	1	1	1
0	1	1	1	1	1	1	1
0	1	0	0	0	1	0	0
0	0	1	0	0	0	1	0
0	0	0	0	0	0	0	0

The fifth column of binary digits of ones and zeros ("on" and "offs") is identical to the list of binary digits in the last column. Therefore, the two circuits are equivalent. It's okay to distribute the "x+' in Boolean algebra..

\- - - - - - - - - - - - -

20. The theory of switching circuits has many applications. Electronic calculators, computers, and telephone systems are designed using the basic concepts of Boolean algebra and switching circuits.

EXAMPLE PROBLEM:

Design a circuit that will turn on a signal light when a majority of a committee of three persons vote yes. Each member of the committee will push a button to vote yes.

SOLUTION:

To record a secret majority vote, the following switching circuit can be used.

Let x, y, z represent the switches or "voting buttons."

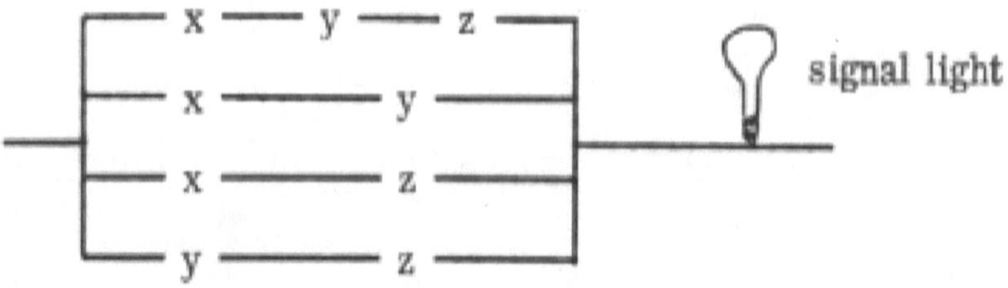

The above circuit will work to indicate a majority vote by turning on the light. However, the above circuit can be simplified. We can design a circuit that employs fewer switches but will be equivalent to the above larger circuit. Simplifying circuits can save money utilizing fewer switches on a circuit board that is part of an electronic device.

One method to simplify circuits in Boolean algebra is to find a simpler equivalent Boolean expression. This can be accomplished by using properties of Boolean algebra.

For example, we know that the following Boolean expression **x + x'** will always produce the same digital output, namely 1. Therefore we can substitute a single wire with no switching required in place of two switches in series that are complementary switches.

$$x + x' = 1$$

In fact, one of Boolean algebra's most important uses is in simplifying digital circuits required for operating digital devices.

21. Let's simplify the circuit we used to record a simple majority vote of three persons. We construct tables for two equivalent Boolean expressions. The second Boolean expression in the second table is simpler and contains 3 less switches. The second circuit with 3 less switches produces the same output as the first circuit table but with less expense .

I. $(x \cdot y) + (x \cdot z) + (y \cdot z) + (x \cdot y \cdot z)$

x	y	z	$(x \cdot y) +$	$(x \cdot z) +$	$(y \cdot z) +$	$(x \cdot y \cdot z)$	value:
1	1	1	1	1	1	1	1
1	1	0	1	0	0	0	1
1	0	1	0	1	0	0	1
1	0	0	0	0	0	0	0
0	1	1	0	0	1	0	1
0	1	0	0	0	0	0	0
0	0	1	0	0	0	0	0
0	0	0	0	0	0	0	0

II. This circuit has 3 less switches: $(x \cdot y) + (x \cdot z) + (y \cdot z)$

x	y	z	$(x \cdot y) +$	$(x \cdot z) +$	$(y \cdot z)$	values are the same:
1	1	1	1	1	1	1
1	1	0	1	0	0	1
1	0	1	0	1	0	1
1	0	0	0	0	0	0
0	1	1	0	0	1	1
0	1	0	0	0	0	0
0	0	1	0	0	0	0
0	0	0	0	0	0	0

Here is a simpler circuit diagram with 3 less switches that lights the signal light when a majority vote is cast.

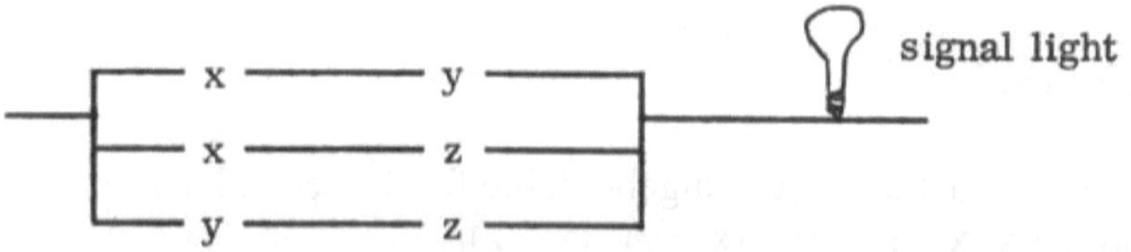

22. EXERCISE 4

1. If two columns representing circuits in a given table are identical, then we can say that their circuits are equivalent.

(a) True (b) False

2. If x = 1, y = 0 and z = 0, then what is the value of the circuit $x \cdot (y + z)$.

(a) 0 (b) 1 (c) none of these

3. The equivalence of the two circuits in the equation $x \cdot (y + z) = x \cdot y + x \cdot z$ is called the first distributive property.

(a) True (b) False

4. The first distributive property is also used in elementary algebra.

(a) True (b) False

5. The equivalence of the two circuits in the equation $x + (y \cdot z) = (x + y) \cdot (y + z)$ is called the second distributive property.

(a) True (b) False

6. The second distributive property is also used in elementary algebra.

(a) True (b) False

7. If x = 1, y = 0 and z = 1 then what is the value of (x • y) + (x • z) + (y • z) + (x • y • z)

(a) 1 (b) 0

8. Construct a Boolean algebra table for the second distributive property: x + (y • z) = (x + y) • (x + z).

23. ANSWERS TO EXERCISE 4

1. (a) True

2. (a) 0

3. (a) True

4. (a) True

5. (a) True

6. (b) False

7. (a) 1

8. The fifth column is the same as the last column, therefore, the second distributive property is true.

x + (y • z) = (x + y) • (x + z)

x	y	z	y • z	x + (y • z)	x + y	x + z	(x + y) • (x + z)
1	1	1	1	1	1	1	1
1	1	0	0	1	1	1	1
1	0	1	0	1	1	1	1
1	0	0	0	1	1	1	1
0	1	1	1	1	1	1	1
0	1	0	0	0	1	0	0
0	0	1	0	0	0	1	0
0	0	0	0	0	0	0	0

24. Next is a summary of properties in Boolean algebra and switching circuits. Some but not all of these properties are found in elementary algebra.

I. Commutative properties of a sum and a product where changing the order of two switches does not change the results. The left side is equivalent to the right side of each of the following Boolean equations.

(a) Commutative property of a sum:

$$x + y = y + x$$

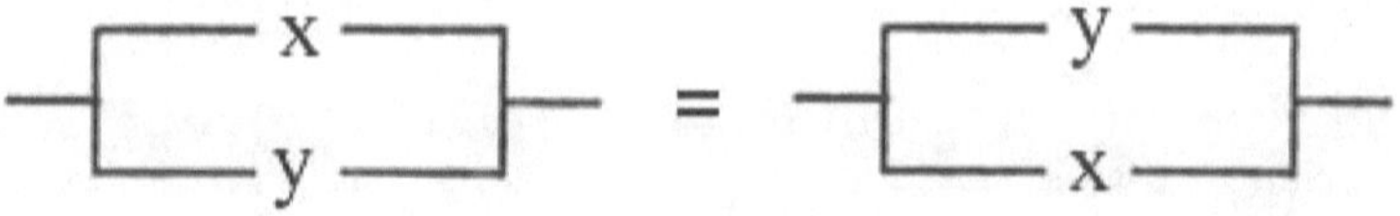

If $x = 1$ and $y = 0$, then $1 + 0 = 0 + 1$

The final result is $1 = 1$

(b) Commutative property of a product:

$$x \cdot y = y \cdot x$$

If $x = 1$ and $y = 0$, then $1 \cdot 0 = 0 \cdot 1$
then the final result is $0 = 0$

II. Associative properties where changing the grouping does not change the results. The left side is equivalent to the right side of each of the following Boolean equations.

(a) <u>Associative property of a sum</u>:

$$(x + y) + z = x + (y + z)$$

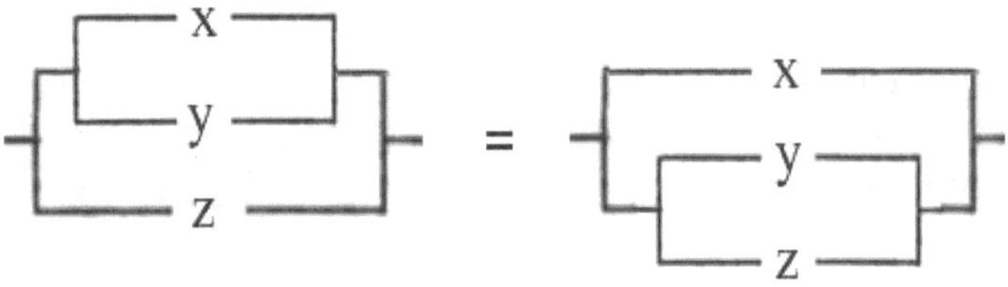

(b) <u>Associative property of a product</u>:

$$(x \cdot y) \cdot z = x \cdot (y \cdot z)$$

—— x —— y —— z —— = —— x —— y —— z ——

III. The identity element for + is 0 and the identity element for $\cdot$ is 1.

$$x + 0 = x$$
$$x \cdot 1 = x$$

IV. The first of the next two other Boolean identity properties is not valid in elementary algebra.

$$x + 1 = 1$$
$$x \cdot 0 = 0$$

V. Idempotent properties:

$x + x = x$ (e.g. $1 + 1 = 1$ and $0 + 0 = 0$)
$x \cdot x = x$ (e.g. $1 \cdot 1 = 1$ and $0 \cdot 0 = 0$)

VI. The following complementary properties were discussed in Section 13.

$x + x' = 1$ (e.g. $1 + 0 = 1$)
$x \cdot x' = 0$ (e.g. $1 \cdot 0 = 0$)

VII. The following distributive properties were discussed in Section 18.

$$x \cdot (y + z) = x \cdot y + x \cdot z$$

$$x + (y \cdot z) = (x + y) \cdot (y + z)$$

VIII. DeMorgan properties. In Exercise 5 you will be asked to verify DeMorgan properties by constructing a table of binary digits for each side of the following equations to verify that they produce or evaluate to an identical set of binary digits.

$$(x + y)' = x' \cdot y'$$

$$(x \cdot y)' = x' + y'$$

25. EXERCISE 5

1. Which property justifies the following Boolean equation?

$$(x + 1) + y = (1 + x) + y$$

(a) commutative property (b) associative property

2. Which property justifies the following Boolean equation.

$$(x + 1) + y = x + (1 + y)$$

(a) commutative property (b) associative property

3. Which property justifies the following Boolean equation?

$$(x \cdot 1) + y = (1 \cdot x) + y$$

(a) commutative property (b) associative property

4. Which properties justifies the following Boolean equation?

$$(x \cdot 1) \cdot y = x \cdot y$$

(a) commutative property (b) Identity property

5. Which property justifies the following Boolean equation.

$$(x + 0) + y = x + y$$

(a) Associative (b) Commutative (c) Identity

6.. Which property justifies the following Boolean equation.

$$(x \cdot 1) + y = x + y$$

(a) Associative (b) Commutative (c) Identity

7. Construct a table verifying: $(x \cdot 1) + y = (x + y)$

8. This time without giving you an answer in the below answer key, construct a table like we did in number 7 to verify that: $(x + 0) + y = x + y$. Each side of this equation on the table should have the same set or order of binary digits.

9. Construct a table for each of the following equations depicting DeMorgan's laws and verify that both sides of each equation are equivalent.

$$(x + y)' = x' \cdot y'$$

$$(x \cdot y)' = x' + y'$$

26. ANSWERS TO EXERCISE 5

1. (a) Commutative property of a sum.

2. (b) Associative property of a sum

3. (a) Commutative property of a product

4. (b) Identity property of a product.

5. (c) Identity property of a sum

6. (c) Identity property of a product

7. The fourth and fifth columns in the following table are the same therefore the two sides of the equation represent equivalent circuits and the following Boolean equation is true:

$$(x \cdot 1) + y = (x + y)$$

x	y	$(x \cdot 1)$	$(x \cdot 1) + y$	$(x + y)$
1	1	1	1	1
1	0	1	1	1
0	1	0	1	1
0	0	0	0	0

8. Each side of the equation in the table you constructed should have the same set and order of binary digits.

9. Verification of DeMorgan's first law which shows that the following Boolean equation is true.:

$$(x + y)' = x' \cdot y'$$

The fourth column (the left side) is equivalent to the last column (the right side). This verifies DeMorgan's second law.

x	y	$x + y$	$(x + y)'$	x'	y'	$x' \cdot y'$
1	1	1	0	0	0	0
1	0	1	0	0	1	0
0	1	1	0	1	0	0
0	0	0	1	1	1	1

----

27. Using properties of Boolean algebra listed in this text, we can justify steps that are used to simplify digital circuits.

EXAMPLES:

(a) Simplify the expression xy + xy' Assume each term is a product like in elementary algebra where the times sign is missing. Name the property used in each step of the simplification process:

$xy + xy'$	$= x\,(y + y')$	First distributive property
	$= x \cdot 1$	Complimentary property
	$= x$	Identity property of a product

(b) Simplify the following expression (x · y) + (x · y') similar to (a) above. In this case the times sign is shown explicitly.

$(x \cdot y) + (x \cdot y')$	$= x \cdot (y + y')$	First distributive property
	$= x \cdot 1$	Complimentary property
	$= x$	Identity property

(c) Prove that the following Boolean equation is true:

$$x \cdot y + x = x$$

In this proof, the below left side is undergoing step-by-step change (justified by Boolean algebra properties) until it appears identical to the expression on the right side.

$x \cdot y + x \cdot 1$	Identity property
$x \cdot (y + 1)$	Distributive property
$x\,(1)$	Identity property
x	Identity property

28. EXERCISE 6

1. Name the property used in each step to prove the following equation's left side is equal to the right side: $(x + x \cdot y) = x$

(a) $x \cdot (1 + y)$

(b) $x \cdot 1$

(c) x

2. Prove that the following Boolean equation is true. Name each property that is used in your proof

$$x \cdot (x + y) = x$$

(3 – 9) Name the Boolean properties used to justify each of the following Boolean statements:

3. $x \cdot x = x$

4. $x + (y + z) = x + (z + y)$

5. $(x + 1) + y = x + (1 + y)$

6. $x = x \cdot 1$

7. $x' \cdot x = 0$

8. $x' + y' = (x \cdot y)'$

9. $1 + x = 1$

10. $x + (y \cdot z) = (x + y) \cdot (x + z)$

29. ANSWERS TO EXERCISE 6

1. (a) First distributive property

   (b) Identity property

   (c) Identity property

2. In the below proof, the left side undergoes step-by-step change (justified by Boolean algebra properties) until it appears identical to the expression on the right side.

$$x \cdot (x + y) = x$$

$x \cdot x + x \cdot y$	Distributive property
$x + x \cdot y$	Idempotent property
$x \cdot (1 + y)$	Distributive property
$x \cdot 1$	Identity property
x	Identity property

3. Idempotent property

4. Commutative property

5. Associative property

6. Identity property

7. Complementary property

8. DeMorgan's property

9. Identity property

10. Second distributive property

This concludes *Introduction to Boolean Algebra and Switching Circuits.*

Other book titles available from amazon.com or whereever books are sold:

We Remember the Day of President Kennedy's Assassination

Handbook for Piano Practice

Piano Practice for the Advancing Student

Windows to Heaven

American Occupation of Korea – One Soldier's Experience

Letters to a Young Classroom Teacher

Program Your Calculator

A Franciscan Odyssey

The Nature Watch Collection Book One

The Nature Watch Collection Book Two

Bird's and Birdwatchers

Political Economics

Made in America

<u>Children's books:</u>

The Calico Caterpillar

Animal Tails

Animal Alphabet Fun Facts

Peppin Puffin to the Rescue

Choices

Time to Fly

Jonah the Reluctant Prophet

Made in United States
Troutdale, OR
11/03/2026

56852548R00056